Machines and Motion

Levers

by Erika S. Manley

Bullfrog Books

Ideas for Parents and Teachers

Bullfrog Books let children practice reading informational text at the earliest reading levels. Repetition, familiar words, and photo labels support early readers.

Before Reading

- Discuss the cover photo. What does it tell them?
- Look at the picture glossary together. Read and discuss the words.

Read the Book

- "Walk" through the book and look at the photos. Let the child ask questions. Point out the photo labels.
- Read the book to the child, or have him or her read independently.

After Reading

- Prompt the child to think more. Ask: Levers are everywhere. Where do you see them? What do they help you accomplish?

Bullfrog Books are published by Jump!
5357 Penn Avenue South
Minneapolis, MN 55419
www.jumplibrary.com

Copyright © 2019 Jump! International copyright reserved in all countries. No part of this book may be reproduced in any form without written permission from the publisher.

Library of Congress Cataloging-in-Publication Data

Names: Manley, Erika S., author.
Title: Levers / by Erika S. Manley.
Description: Minneapolis, MN : Jump!, Inc., [2018]
Series: Machines and motion
"Bullfrog Books are published by Jump!"
Audience: Ages 5–8. | Audience: K to grade 3.
Includes bibliographical references and index.
Identifiers: LCCN 2017051887 (print)
LCCN 2017052898 (ebook)
ISBN 9781624968532 (e-book)
ISBN 9781624968518 (hardcover : alk. paper)
ISBN 9781624968525 (pbk.)
Subjects: LCSH: Levers—Juvenile literature.
Simple machines—Juvenile literature.
Classification: LCC TJ147 (ebook)
LCC TJ147 .M22694 2018 (print) | DDC 621.8/11—dc23
LC record available at https://lccn.loc.gov/2017051887

Editor: Kristine Spanier
Book Designer: Molly Ballanger

Photo Credits: Sergiy Kuzmin/Shutterstock, cover; El Greco/Shutterstock, 1; Margo Harrison/Shutterstock, 3; BLUR LIFE 1975/Shutterstock, 4 (background); KK Tan/Shutterstock, 4 (foreground); Andreis83/Shutterstock, 5 (lever), 23br; Tushchakorn/Shutterstock, 5 (coins), 23br; Syda Productions/Shutterstock, 6–7, 23bl; Andrey _ Popov/Shutterstock, 8–9, 23tl; Nordroden/iStock, 10–11; Planner/Shutterstock, 12, 23mr; Image Source/Shutterstock, 13; olesiabilkei/iStock, 14–15, 23ml, 23tr; aleks333/Shutterstock, 16–17 (background); Clari Massimiliano/Shutterstock, 16–17 (foreground); Ronnie Chua/Shutterstock, 18; Steve Skjold/Alamy, 19; Steve Debenport/iStock, 20–21; Cheryl Ann Quigley/Shutterstock, 24.

Printed in China.

Table of Contents

Lifting with Levers

What is a lever?

It is a simple machine.
It helps us lift things.
It helps us move them. How?

Four parts make it work.
We press on the lever.
Leo pulls out a nail.

lever

load
effort
fulcrum

We use effort.

The fulcrum holds the load.

We lift it.

Levers make work easier.

lever

A pliers is a lever.

Bo changes a tire.

A seesaw is a lever.

Kay sits on one end.

The other end
goes up.

fulcrum

Ali opens a bottle.

She uses an opener.

She pulls up.

The cap comes off.

A hockey stick is a lever.

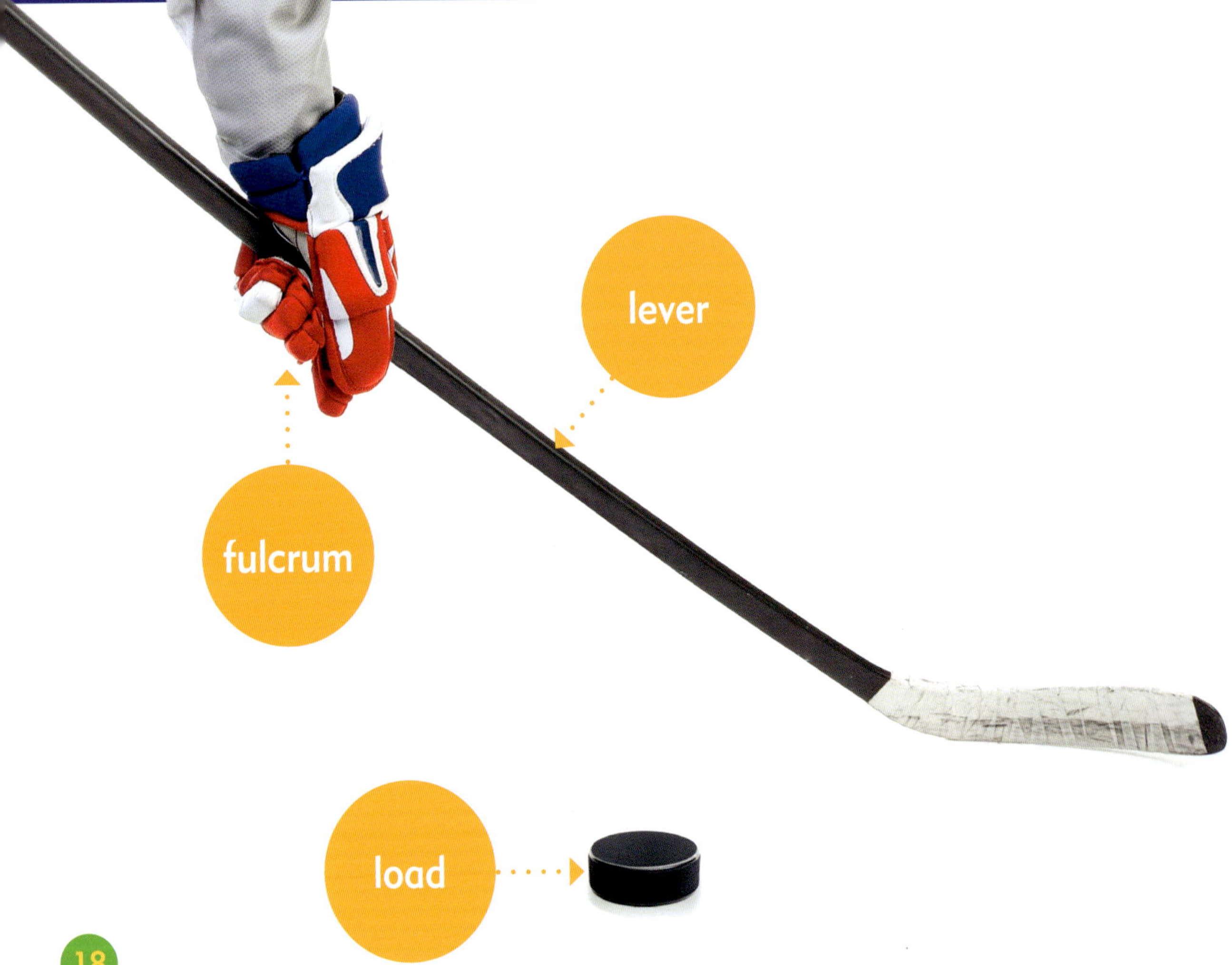

Logan hits the puck.

Levers help us!

Build a Lever

Levers help us move or lift things. Build your own lever!

You will need:

- ruler
- flat block eraser
- cotton ball

Directions:

1. Place the eraser on a flat surface.
2. Lay the center of the ruler on the eraser.
3. Place the cotton ball on one end of the ruler. Press the other end of the ruler down. Notice how much effort it takes to raise the cotton ball. You just made a lever!
4. There are four parts that make a lever work. The lever itself, the fulcrum, the load, and effort. What part is the eraser? What part is the ruler?

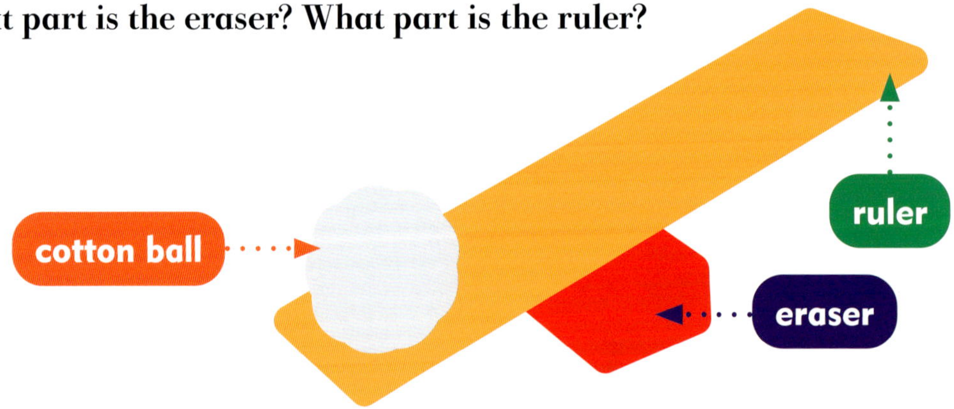

Picture Glossary

effort
The force used to move or lift weight.

load
The weight being lifted or moved.

fulcrum
The point on which a lever rests or turns.

pliers
A tool with two handles and jaws that can grip and bend or cut objects.

lever
A bar or handle that rests or pivots to help lift or move weight.

simple machine
A tool used to make work easier, such as an inclined plane, lever, pulley, screw, wedge, or wheel and axle.

Index

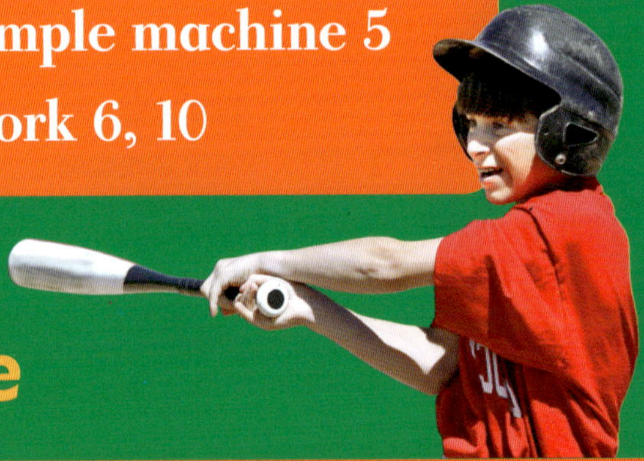

To Learn More

Learning more is as easy as 1, 2, 3.

1) Go to www.factsurfer.com
2) Enter "levers" into the search box.
3) Click the "Surf" button to see a list of websites.

With factsurfer.com, finding more information is just a click away.